HORSES

BOLD & BEAUTIFUL

Publications International, Ltd.

WRITTEN BY BETH TAYLOR

Images from the Library of Congress, the National Park Service, Shutterstock.com and Wikimedia Commons

Louis Weber, CEO
Publications International, Ltd.
8140 Lehigh Avenue
Morton Grove, IL 60053

ISBN: 978-1-64030-737-7

Manufactured in China.

8 7 6 5 4 3 2 1

CONTENTS

INTRODUCTION

Humans have a special relationship with horses that dates back millennia. Since they were first domesticated about 6,000 years ago, horses have plowed fields and brought in the harvest, hauled goods and transported passengers, followed game and tracked cattle, carried combatants into war and explorers to unknown lands.

The height of a horse is measured in hands, from the ground to the top of the withers, the ridge between the horse's shoulder bones at the base of the neck. One hand is equal to four inches.

LIFE STAGES

Young horses less than one year of age are foals. Yearlings are between one and two years old. Colts are male horses less than four years old; fillies are females under age four. A mature male horse is called a stallion; the female is a mare. A stallion used for breeding is known as a stud. A castrated male horse is called a gelding.

COAT COLORS

A horse's coat color is determined by its genetic makeup, which is inherited equal from its sire (father) and dam (mother). Common colors include bay (reddish-brown), brown, gray (black skin with white or mixed dark and white hairs), dun (yellowish-tan), chestnut (reddish-gold), sorrel (light chestnut), palomino (golden), roan (white intermixed with another color), pinto (spotted), and black.

HORSE BREEDS

More than 300 breeds of horses and ponies exist around the world. Horse breeds can be classified in many ways, including where they originated (e.g., Arabian, Percheron, Icelandic), their primary use (riding, draft, stock, coach horse), their temperament (cold blood, warmblood, hot blood), and their outward appearance and size (light, heavy, pony).

Shire horses are large draft horses known for their capacity to pull heavy loads. Many measure more than 16 hands, or 64 inches, high at the withers.

DRAFT HORSES

Draft (or draught) horses were bred to carry and pull heavy loads. They are tall, heavy, and strong. They are described as "cold bloods" based on their calm, patient, and levelheaded temperament. The average draft horse stands over 16 hands (64 inches) tall and weighs over 1,600 pounds. Before the age of machines, draft horses were used for plowing fields, logging, hauling heavy loads, and other tasks that required great pulling capacity. Draft breeds include the Clydesdale, Percheron, Belgian, and Shire horses.

LIGHT HORSES

Light horses were bred for speed, agility, endurance, and riding. Light horses are used for every type of riding, including pleasure riding, racing, and ranch work. Light horse breeds tend to be more spirited, energetic, and easily excitable than their heavy draft breed counterparts. Many light horse breeds—especially Thoroughbreds—are called "hot bloods" based on their high-strung temperament. Light horse breeds include the Arabian, Morgan, and American quarter horses.

WARMBLOODS

Warmbloods are intermediate-weight horses created by crossing draft horses (cold bloods) and light horses (hot bloods). These horses are usually calmer than light horse breeds, and more athletic and agile than heavy draft breeds. Warmbloods dominate dressage, jumping, and other equestrian sports. Warmblood breeds include the Dutch warmblood, Hanoverian, Trakehner, Oldenburg, and Holsteiner.

PONIES

Ponies are horses shorter than 14.2 hands (56.8 inches). Ponies are versatile, sturdy, and sometimes-stubborn little horses valued for their intelligence, strength, and hardiness. Ponies are used for everything from pulling carriages and pack loads to pets and riding horses for children. Pony breeds include the Welsh, Connemara, New Forest, Highland, Shetland, and Fell pony.

ARABIAN

CHAPTER 1:

Horses of Asia

The Arabian is one of the world's oldest horse breeds. For thousands of years, Arabians lived among the desert tribes of the Arabian Peninsula, used by the Bedouins for transportation on long treks, beasts of burden, and war mounts. Its striking beauty, speed, stamina, intelligence, and gentleness have made the Arabian a popular breeding horse throughout the ages. Today Arabian bloodlines are found in nearly every light horse breed.

The Arabian developed in a desert climate and was prized by the nomadic Bedouin people. Arabians were often brought inside the family tent at night to shelter them from the cold and protect them from theft.

Arabian horses typically have a small head, with a dished (concave) profile, wide-set eyes, and large, efficient nostrils.

BREED	Arabian
ORIGIN	Arabian Peninsula
COLOR	Chestnut, gray, bay, black, roan
HEIGHT	14.1–15.1 hands (56.4–60.4 inches)
WEIGHT	800–1,000 pounds

Arabians are compact, relatively small horses with a finely chiseled head, prominent eyes, wide nostrils, sloped shoulders, an arched neck, and a short back. Arabian horses have one fewer vertebra than other horse breeds. They have fine, silky coats and black skin.

AKHAL-TEKE

The Akhal-Teke descends from the raiding horses of Turkmenistan. The breed's name comes from the Akhal oasis in the Kopet Dag Mountains and the nomadic Teke tribe who inhabited the area and bred the horses. The Akhal-Teke horse is known for its speed, endurance, and intelligence.

Akhal-Tekes are often used for dressage, show jumping, and long-distance racing.

The Akhal-Teke's sleek, athletic body with its long, narrow chest is often compared to a cheetah or greyhound.

BREED	**Akhal-Teke**
ORIGIN	**Turkmenistan**
COLOR	**Bay, black, chestnut, gray, palomino, cremello, perlino**
HEIGHT	**14.2–16 hands (56.8–64 inches)**
WEIGHT	**900–1,000 pounds**

Akhal-Tekes are noted for the distinctive metallic sheen of some individuals. Many carry a cream dilution gene, which can result in perlino, palomino, and cremello coat colors.

PRZEWALSKI'S HORSE

The Przewalski's horse, also known as the Mongolian wild horse or the takhi, is a rare and endangered horse native to the steppes of central Asia. Until the late 18th century, these wild horses ranged from the Russian Steppes east to Kazakhstan, Mongolia, and northern China. The Przewalski's horse had become extinct in the wild by the 1960s, but has since been successfully reintroduced into its native habitat in Mongolia.

A group of Przewalski's horses was introduced into the Chernobyl Exclusion Zone in 1998 and has grown in the years since. The area has been protected from human interference since the 1986 nuclear accident and now serves as a de facto nature reserve.

The Przewalski's horse is a small, stocky wild horse with a large head, a short, upright mane, and a low-set tail.

Khustain Nuruu National Park, also known as Hustai National Park, is one of the sites in Mongolia where Przewalski's horses have been reintroduced.

All Przewalski's horses alive today are descended from 12 wild-caught individuals, and as many as four domestic horses.

KYRGYZ

The Kyrgyz horse is strongly associated with the Kyrgyz people's nomadic past. It remains an important part of Kyrgyz culture and national identity today.

The Kyrgyz horse has been bred in the steppes of central Asia for thousands of years. These small, hardy horses are well suited to the harsh terrain in the Tian Shan Mountains and essential to the nomadic way of life. The Kyrgyz horse is used for everything from transportation and traditional mounted games to meat and milk production.

BREED	**Kyrgyz**
ORIGIN	**Kyrgyzstan**
COLOR	**Bay, chestnut, gray, roan, buckskin**
HEIGHT	**13.2–14.2 hands (52.8–56.8 inches)**

Milk from Kyrgyz mares is often fermented to make a drink called kumis.

Er enish, or oodarysh, is a traditional Kyrgyz equestrian sport. This form of wrestling from horseback is played at the World Nomad Games.

MARWARI

The average Marwari stands between 14.2 and 15.2 hands (56.8–60.8 inches) tall.

The Marwari is a rare horse breed from the Marwar (or Jodhpur) region of northwestern India. For centuries the Marwari has been used by the region's people as a cavalry mount. The Marwari is a hardy breed well suited to the desert environment.

A wide variety of colors have been observed in Marwari horses. Common coat colors include gray, white, bay, chestnut, dun, fleabitten gray, skewbald, and piebald.

The Marwari's inward-curling ears are one of its most distinctive features.

Zanskari ponies are known for their ruggedness and often used as pack animals. They work better than some other breeds at high altitudes.

ZANSKARI

The Zanskari, or Zaniskari, pony is a small mountain horse native to northern India. The breed gets its name from the Zanskar valley. The Zanskari is considered endangered, as only a few hundred of the ponies exist today.

Gray is the most common color of the Zanskari pony. Other common colors include bay, brown, black, and chestnut.

Zanskari pony

KAZAKH

The Kazakh is a horse breed of the Kazakh people, who live in Kazakhstan and parts of Mongolia, China, Russia, and Uzbekistan. Kazakh horses are mainly used for riding, but also for meat and milk production. The breed is known for its hardiness and stamina.

Kazakh eagle hunters used trained golden eagles to hunt foxes from horseback. This traditional form of falconry is practiced in Kazakhstan, Kyrgyzstan, and Mongolia.

MONGOL

The Mongol horse is native to Mongolia. These short, stocky horses live outdoors all year long, surviving brutal winter temperatures. Mongol horses are used for riding, carting supplies, milk, and meat production.

Mongol horse

MISAKI

The Misaki is an endangered small horse breed native to Japan. The Misaki horses and the area they inhabit, Cape Toi, were declared a natural monument. Cape Toi is located along the southeastern edge of Miyazaki Prefecture on the Japanese island of Kyushu.

The most common colors for Misaki horses are black, bay, and chestnut.

Misaki mare and foal

The Misaki horses are a popular draw for tourists.

SABLE ISLAND HORSE

CHAPTER 2:
Horses of North America

Sable Island horses are named for the narrow island they inhabit off the coast of Nova Scotia, Canada. The short, stocky horses are the only land mammals on Sable Island, aside from the few human inhabitants. Federal law protects these small feral horses from human interference.

Sable Island horses have thick shaggy coats, manes, and tails that help them survive harsh winters.

BREED	**Sable Island horse**
ORIGIN	**Sable Island, Canada**
COLOR	**Bay, chestnut, palomino, black**
HEIGHT	**13–14 hands (52–56 inches)**
WEIGHT	**660–790 pounds**

Sable Island mare and foal

Sable Island horses eat marram grass, which grows on coastal sand dunes.

CANADIAN HORSE

The Canadian horse descended from draft and light riding horses brought from France to the colonies of Acadia and New France in the late 1600s. It was used for plowing fields and pulling carriages. The Canadian horse was used to improve the strength and hardiness of other breeds. It helped found other North American breeds, including the Morgan, Tennessee walking horse, and American saddlebred. Because of its strength, endurance, and ability to survive harsh weather, scarce food, and vigorous work, the Canadian horse earned the nickname "the little iron horse."

BREED	**Canadian horse**
ORIGIN	**Canada**
COLOR	**Black, bay, brown, chestnut**
HEIGHT	**14–16 hands (56–64 inches)**
WEIGHT	**1,000–1,350 pounds**

NEWFOUNDLAND PONY

The Newfoundland pony traces its ancestry to the Exmoor, Dartmoor, New Forest, Galloway (now extinct), Welsh, Connemara, and Highland ponies brought by early settlers. For three centuries, these ponies interbred until the Newfoundland pony developed. This pony is perfectly suited for the rugged Newfoundland environment.

The Newfoundland pony can survive harsh winters due to its thick coat and mane.

Canadian horses are characterized by their strong, sturdy legs; arched necks; tough feet; long, wavy manes and tails; and finely chiseled heads. In addition to their hardiness, Canadian horses are also known for their even temperament and willingness to please.

BREED	**Newfoundland pony**
ORIGIN	**Newfoundland, Canada**
COLOR	**Bay, black, brown, chestnut, dun, gray, roan, white**
HEIGHT	**11-14.2 hands (44-56.8 inches)**
WEIGHT	**400-800 pounds**

MUSTANG

American mustangs come in all possible colors.

The American mustang is a feral horse found in the western United States. The English word *mustang* comes from the Spanish word *mesteño*, meaning wild, stray, or ownerless. The mustang first descended from domesticated horses the Spanish brought to the Americas but eventually became a mix of numerous breeds. Today the Bureau of Land Management (BLM) protects and manages the free-roaming mustang population.

Significant populations of free-roaming mustangs exist in Nevada, Wyoming, California, Oregon, Utah, and Montana. These mustangs were photographed in Wyoming.

American mustangs in the Sierra Nevada

PRYOR MOUNTAIN MUSTANG

The Pryor Mountain mustang is a strain descended from the colonial Spanish horses brought to North America in the mid-1600s. They live on the Pryor Mountains Wild Horse Range, a protected refuge for feral mustangs, located on the Montana-Wyoming border.

Kiger mustangs stand about 13–15 hands (52–60 inches) tall and weigh about 750–1,000 pounds. They come in all colors, although dun is the most common.

Pryor Mountain mustangs

KIGER MUSTANG

Kiger mustangs are genetically related to the Spanish horses brought to North America in the 1600s. This bloodline was thought to have largely disappeared from mustang herds before the Kiger mustangs were found. Kiger mustangs are known for their intelligence and stamina.

SPANISH MUSTANG

The Spanish mustang is an American horse breed descended from horses brought from Spain during the early conquest of the Americas. Noted for its stamina and toughness, this breed truly shows its Spanish inheritance in its ability to survive where other breeds would have perished.

HORSES OF THE BADLANDS

Feral horses have existed in the Badlands of North Dakota since the mid-1800s. With the development of modern ranching, feral horses came to be regarded as a nuisance that competed with livestock and depleted the range. Cattle ranchers worked to exterminate these animals throughout the West. Efforts to conserve feral horses began in the 1950s and 1960s. In 1971, the Wild Free-Roaming Horses and Burros Act mandated the protection of these feral horses as a "national heritage species." Today Theodore Roosevelt National Park in North Dakota is home to numerous bands of feral horses.

Theodore Roosevelt National Park is one of the few areas in the West where free-roaming feral horses can be observed.

Feral horses typically range in small bands of 5–15 animals. Each group has an established social hierarchy, consisting of a dominant stallion, his mares, and their offspring.

Feral horses in Theodore Roosevelt National Park

Feral horses in Theodore Roosevelt National Park show some interesting color characteristics, including many with a white or "bald" face and spots or "dappling."

Young horses stay with their birth herd until they are two or three years old, at which time they leave and form new bands. Below are two young horses with their mother.

Solitary horses tend to be bachelor stallions that are trying to form a herd of their own.

NOKOTA HORSE

The blue roan coat color is uncommon in other horse breeds, but common in Nokota horses.

Band of Nokota horses

The Nokota horse developed in the Badlands of southwestern North Dakota. These wild horses have inhabited the Little Missouri Badlands, now encompassed by Theodore Roosevelt National Park, for more than a century.

MORGAN

The Morgan breed originated with a stallion named Figure given to Justin Morgan of Vermont in the late 1700s. Though the horse died in 1821, his influence still persists. The Morgan has contributed qualities to other major breeds, including the Tennessee walking horse, American standardbred, and American quarter horse.

The Morgan horse is known for its versatility. Morgans have been used for general riding, racing, pulling carriages, and farm labor. Morgans served as cavalry horses on both sides of the American Civil War. Today Morgans are used in a number of English and Western equestrian events, including dressage and show jumping.

Shown here is a Morgan at the University of Vermont's Morgan horse farm. The Morgan is the state animal of Vermont.

The Morgan is a compact, robust, good-natured, and intelligent breed.

BREED	Morgan
ORIGIN	United States
COLOR	Bay, black, brown, chestnut, gray, roan, dun, palomino, cremello, perlino, buckskin
HEIGHT	14.1–15.2 hands (56.4–60.8 inches)
WEIGHT	900–1,100 pounds

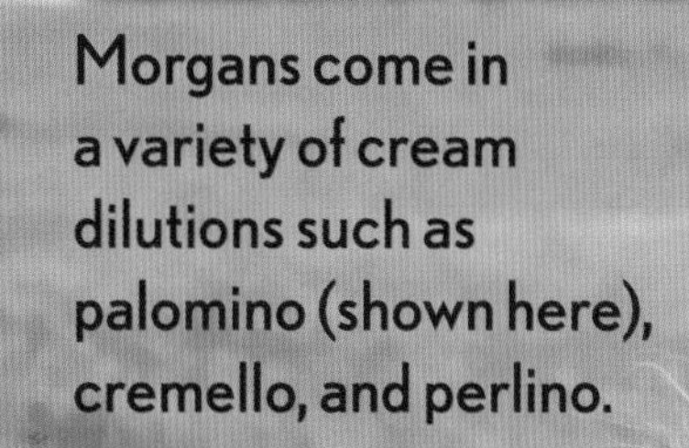

Morgans come in a variety of cream dilutions such as palomino (shown here), cremello, and perlino.

AMERICAN QUARTER

The American quarter horse originated in the 17th century when colonists crossed imported English Thoroughbreds with Native American horses of Spanish origin. The result was a compact, muscular horse perfectly suited for the short-distance races colonists loved. The breed's name actually comes from the quarter-mile race at which it excels. The quarter horse is also known for its "cow sense"—a natural instinct for working with cattle. Today quarter horses are used for everything from racing and ranching to rodeo events and recreational riding.

The American quarter horse is the most popular breed in North America.

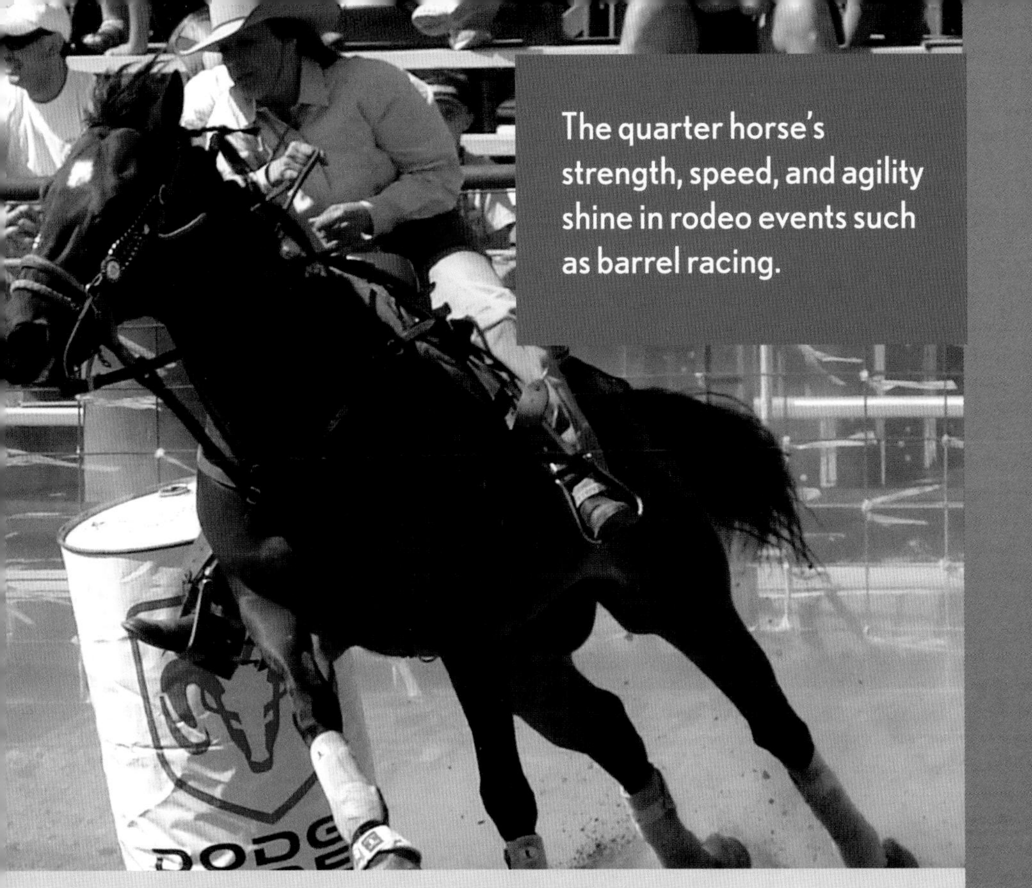

The quarter horse's strength, speed, and agility shine in rodeo events such as barrel racing.

BREED	**American quarter**
ORIGIN	**United States**
COLOR	**Sorrel, bay, black, brown, buckskin, chestnut, dun, red dun, gray, grullo, palomino, red roan, blue roan**
HEIGHT	**14.3–16 hands (57.2–64 inches)**
WEIGHT	**900–1,200 pounds**

The quarter horse is surefooted and nimble, even when racing short distances at great speeds. Some have been clocked at speeds up to 55 miles per hour.

STANDARDBRED

The standardbred is an American breed developed in the 19th century and used for harness racing. An English Thoroughbred named Messenger is considered the foundation sire of the breed. His progeny were crossed with other breeds, including the Morgan, Canadian pacer, and the now-extinct Norfolk Trotter, which contributed desirable racing characteristics. The breed's name comes from the fact that horses were required to reach a certain "standard" timed speed at the mile in order to be registered as part of the breed.

BREED	**Standardbred**
ORIGIN	**United States**
COLOR	**Bay, brown, black, chestnut, gray**
HEIGHT	**15–16 hands (60-64 inches)**
WEIGHT	**900-1,200 pounds**

The standardbred is best known as a harness racing breed.

AMERICAN CREAM DRAFT

The American cream is the only draft horse breed developed in the United States still in existence today. The breed is known for its cream color, known as "gold champagne." The American cream draft breed descends from a foundation mare named Old Granny from Iowa.

The ideal American cream draft horse has a medium cream color with a white mane and tail, pink skin, and amber eyes.

AMERICAN SADDLEBRED

The American saddlebred horse, also called the American saddle horse, is the dominant riding horse used in horse shows in the United States. It counts the Thoroughbred, Morgan, standardbred, and Canadian pacer among its ancestors.

AMERICAN PAINT HORSE

Paint horses are typically very muscular, with broad chests and strong hindquarters.

The American paint horse combines the characteristics of a Western stock horse with the distinctive pinto coat pattern. The breed's descendants arrived in North America with Spanish conquistadors. Paint horses became part of the wild herds that roamed the West. Today paint horses are used in nearly every Western and English equestrian discipline.

Tobiano is one of the main color patterns paint horses can have. Tobianos typically have white legs and a dark head with regular facial patterns such as stars, blazes, and strips. The tail and mane can be of two colors.

Various breeds can have a pinto coat pattern (including the pony above), but the paint horse is a separate breed.

The paint horse's coat patterns can occur in any combination of white and another color, such as bay, black, palomino, or chestnut. Some paint horses are a solid or almost-solid color.

MOUNTAIN PLEASURE HORSE

Mountain pleasure mare and foal

The mountain pleasure horse was developed in the Appalachian Mountains of eastern Kentucky. The breed has existed in the area for more than 160 years. The mountain pleasure horse contributed to the development of other breeds, including the Tennessee walking horse, American saddlebred, and more recently, the Rocky Mountain horse. Mountain pleasure horses are valued for their smooth gaits, hardiness, and calm temperaments.

TENNESSEE WALKING HORSE

Palomino Tennessee walking horse

Black Allan (later known as Allan F-1) is considered the foundation sire of the Tennessee walking horse breed.

The Tennessee walking horse is one of North America's most popular horse breeds. The breed gets its name from the state where it originated and its distinctive gait—the running walk. Originally used for farm and plantation work, the Tennessee walking horse is now primarily a riding horse equally prized in the show ring and on the trail.

KENTUCKY MOUNTAIN SADDLE

Kentucky mountain saddle horses are prized for their calm temperament, intelligence, versatility, willingness, and smooth gait.

The Kentucky mountain saddle horse was developed in eastern Kentucky as a farm and riding horse. The breed is related to the Tennessee walking horse and other gaited breeds. Kentucky mountain saddle horses are known for their gentle temperament and willing disposition.

MISSOURI FOX TROTTER

Missouri fox trotter mare

Missouri fox trotters may be any solid color or pinto. White markings on the face and legs are common.

Missouri fox trotters are known for their distinctive diagonal fox trot gait.

The Missouri fox trotter was developed in the Ozark Mountains in the 19th century. They are valued for their stamina and smooth gaits. The breed performs an ambling gait known as the "fox trot," a four-beat broken diagonal gait in which the horse appears to walk with its front legs and trot with its hind legs.

APPALOOSA

Spotted horses have existed for millennia. Cave paintings dating to 25,000 years ago depict spotted horses. The Appaloosa is one such breed of spotted horse. The Nez Percé people developed the Appaloosa horse in North America. The breed's name likely derives from the Palouse River of Idaho and Washington. Appaloosas have several distinct coat patterns. Other distinguishing characteristics are mottled skin, white sclera, and striped hooves. The modern Appaloosa horse is used for pleasure and trail riding, working cattle and rodeo events, racing, and many other Western and English riding disciplines.

Appaloosas can have a mainly white body with dark leopard spots.

Many Appaloosas, like the one below, have a white section interspersed with colored dots covering the hip area. This coat pattern is called "blanket with spots."

The Appaloosa is an American breed best known for its spotted coat patterns.

Readily visible white sclera (the portion of the eye surrounding the iris) and mottled skin (speckled or blotchy pattern of pigmented and non-pigmented skin) are two of the identifiable characteristics of the Appaloosa breed.

Mottled skin is commonly seen around the Appaloosa's muzzle and eyes, like in the photo below.

Readily visible white sclera is a distinctive characteristic of Appaloosas. Most other breeds can only show white around the eye when the eye is rolled back.

Appaloosas typically have alternating dark and light stripes running vertically on their hooves.

Nez Percé Indians with Appaloosa horse circa 1895

PONY OF THE AMERICAS

The Pony of the Americas horse gets its unique coloring from the Appaloosa. The POA typically also has mottled skin, visible white sclera, and striped hooves.

The Pony of the Americas (POA) is a riding-pony breed used as a child's mount. The breed was developed in the state of Iowa in the 1950s by crossing ponies with Appaloosa horses. In order to be registered with the Pony of the Americas Club, a horse must have the Appaloosa patterning and measure from 11.5 to 14 hands (46–56 inches) tall at maturity. The Pony of the Americas is known for its gentle disposition, durability, and intelligence.

ASSATEAGUE HORSE

Off the coasts of Maryland and Virginia is the 37-mile-long barrier island of Assateague, which is home to feral horses alternately known as Assateague horses in Maryland and Chincoteague ponies in Virginia. These small but sturdy horses have adapted to survive the scorching heat, abundant insects, stormy weather, and poor quality food found on the windswept barrier island. Assateague's horses are split into two herds, separated by a fence along the Maryland–Virginia State line. The National Park Service manages the Maryland herd.

Assateague's horses primarily eat saltmarsh cordgrass, saltmeadow hay, and beach grass. To compensate for their salty diet, they drink twice the amount of water as domesticated horses. This contributes to their bloated appearance.

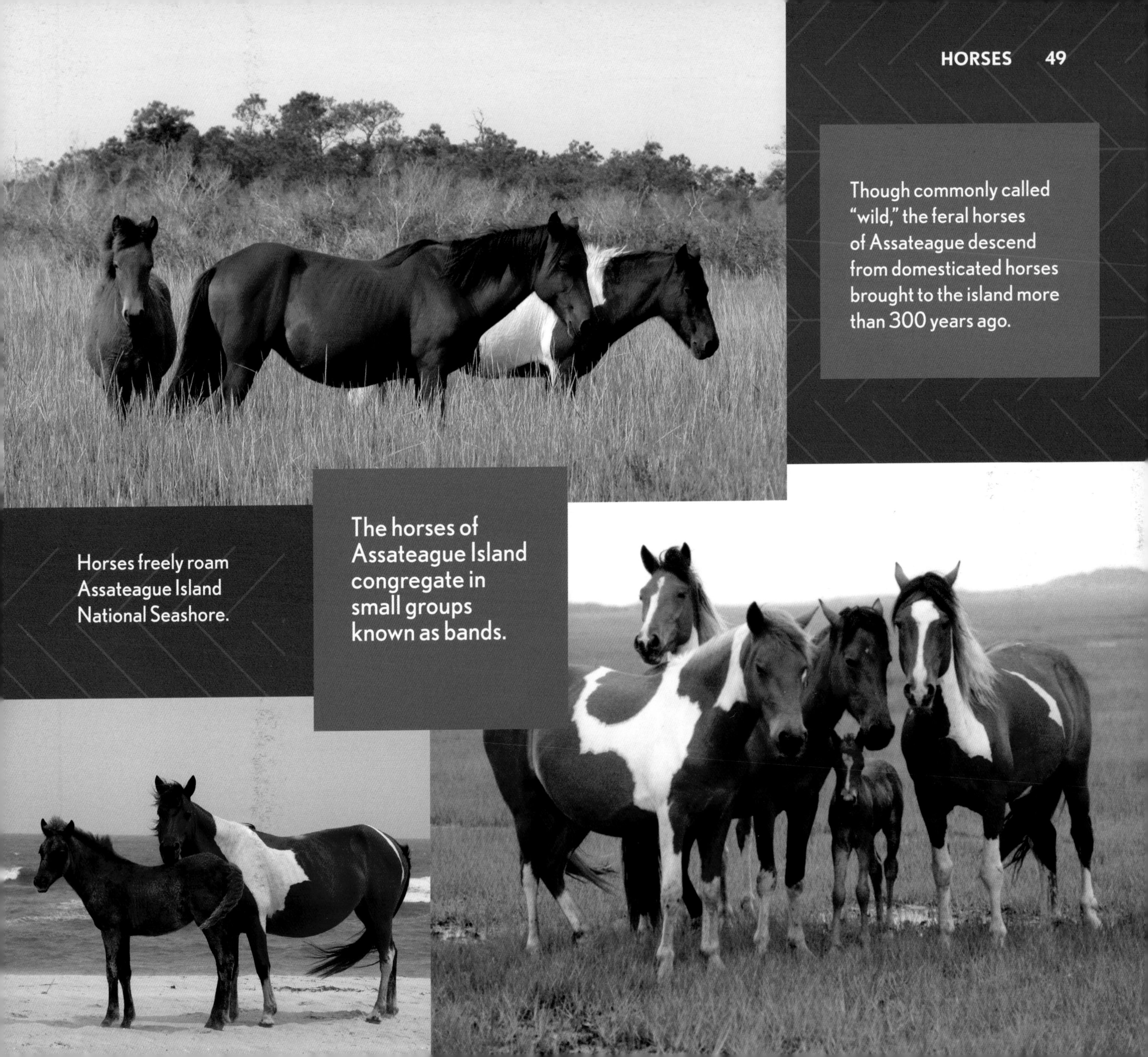

Though commonly called "wild," the feral horses of Assateague descend from domesticated horses brought to the island more than 300 years ago.

Horses freely roam Assateague Island National Seashore.

The horses of Assateague Island congregate in small groups known as bands.

CHINCOTEAGUE PONY

Chincoteague ponies, also known as Assateague horses, inhabit the Virginia side of Assateague Island. The Virginia herd is owned and managed by the Chincoteague Volunteer Fire Company. Chincoteague ponies are stocky, with short legs, thick manes, and large, round bellies. A local legend claims the island's horses descend from shipwreck survivors who swam ashore. The more plausible explanation is that they descend from horses brought to Assateague Island by mainland owners in the late 1600s to avoid fencing laws and livestock taxes.

Chincoteague ponies are found in various solid colors and pinto coat patterns.

Horses in the Virginia herd are allowed to graze in the Chincoteague National Wildlife Refuge with a special permit. Annual auctions keep the Virginia population at levels required by agreement with the Refuge.

Kayaker passing Chincoteague ponies on Assateague Island

Every year, the Virginia herd is rounded up and swum from Assateague Island to nearby Chincoteague Island. Most of the young foals are auctioned the following day. This helps control the size of the herd and raises funds for the Chincoteague Volunteer Fire Company.

AMERICAN MINIATURE

American miniature mare and foal

The American miniature is a height breed; these horses must measure no more than 34 inches from the last hairs of the mane (at the withers) to the ground. These tiny equine are scaled-down versions of larger horse breeds. In addition to their small size, American miniatures are known for their gentle nature, curiosity, and intelligence.

Miniature horses are often used as pets or companions for young children, adults, seniors, and people with disabilities.

American miniatures have a wide forehead and large, deep-set eyes.

Miniature horses come in every possible color and coat pattern.

American miniature foals measure from 16 to 21 inches in height at birth.

In order to be registered with the American Miniature Horse Association, like this horse, American miniatures must be no taller than 34 inches.

American miniatures are used for various equestrian competitions, including jumping, driving, obstacle running, costume, and showmanship.

PASO FINO

The Paso Fino is prized for its smooth gait.

Puerto Rican Paso Fino stallion

The Paso Fino was developed in the Caribbean by crossbreeding Andalusian, Barb, and Jennet horses of Spanish stock. These small, agile horses helped in the conquest, exploration, and development of the Americas. The Paso Fino breed is known for its smooth, natural, four-beat, lateral gait.

CUMBERLAND ISLAND HORSE

Feral horses at Cumberland Island National Seashore

Cumberland Island is Georgia's largest and southernmost barrier island. It's home to some 150 feral horses that have descended from modern, domestic breeds. Genetic studies by the University of Georgia and University of Kentucky showed that Cumberland Island's horses are closely related to Tennessee walking horses, American quarter horses, Arabians, and Paso Finos.

Mare and foal on Cumberland Island

Cumberland Island horses are often seen grazing in the salt marsh, dune meadows, unplanted fields, and among historic ruins.

BANKER HORSE

Shackleford Banks is home to more than 100 feral banker horses.

The banker horse is a breed of feral horse living in North Carolina's Outer Banks. Ocracoke Island, Shackleford Banks, Currituck Banks, and the Rachel Carson Estuarine Sanctuary have populations of banker horses. The National Park Service, the state of North Carolina, and several private organizations protect the horses.

Feral banker stallion at Cape Lookout National Seashore

Banker horses survive by grazing on marsh grasses.

Banker foal

PERUVIAN PASO

CHAPTER 3:

Horses of South America

The Peruvian Paso, sometimes called the Peruvian horse, is a light saddle horse known for its smooth gait, strength, stamina, willing attitude, and gentle demeanor. Peruvian Pasos descend from Spanish stock horses brought to South America in the 1500s. The breed developed in Peru by blending the Spanish Jennet's smooth ambling gait and even temperament; the Barb's stamina, strength, and energy; and the Andalusian's conformation, action, and beauty. For centuries, breeders in Peru kept Peruvian Paso bloodlines pure and selectively bred horses for gait, temperament, and conformation. Today this versatile horse is used for pleasure and trail riding, parades, and horse shows.

Shown here are Peruvian Paso riders in traditional clothing in Trujillo, Peru. Peru's government protects the Peruvian Paso horse as part of the country's cultural heritage.

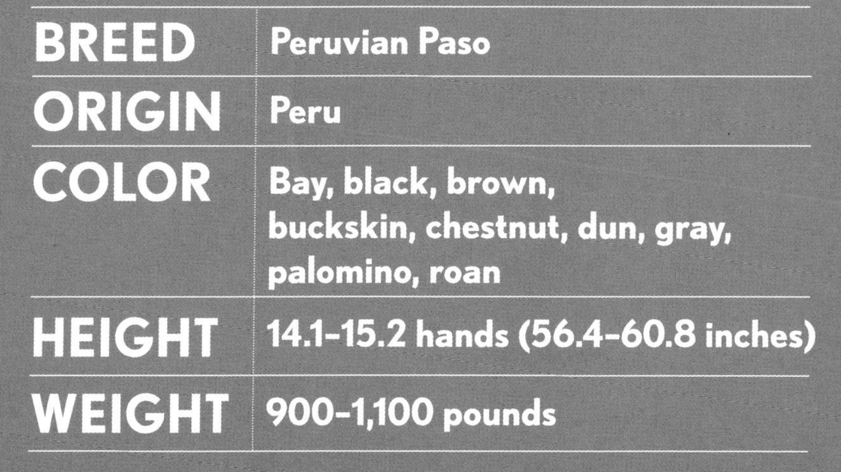

BREED	Peruvian Paso
ORIGIN	Peru
COLOR	Bay, black, brown, buckskin, chestnut, dun, gray, palomino, roan
HEIGHT	14.1–15.2 hands (56.4–60.8 inches)
WEIGHT	900–1,100 pounds

A unique feature of the Peruvian Paso is that all purebred foals inherit the breed's characteristic smooth gait.

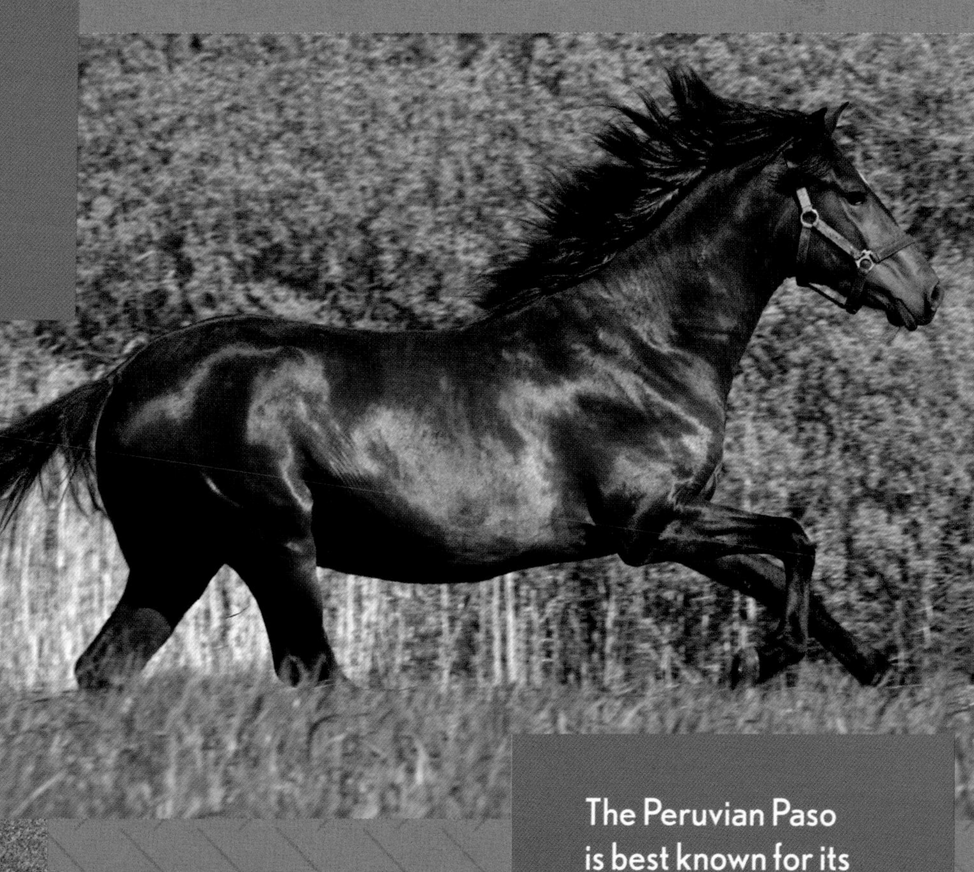

The Peruvian Paso is best known for its easy gait.

FALABELLA

Falabellas come in a wide variety of coat colors. Brown, bay, black, and chestnut are the most common.

The Falabella miniature horse is one of the smallest horse breeds in the world. A mature Falabella typically stands between 28 and 34 inches tall at the withers. However, there is no breed standard for height. Despite its size, the Falabella is considered a miniature horse rather than a pony. The breed gets its name from the family that developed the breed in Argentina in the mid-19th century.

Falabella foals are only 12–22 inches tall at birth.

ARGENTINE CRIOLLO

Dun is the most popular color, but Criollos also come in bay, brown, black, chestnut, grullo, buckskin, palomino, roan, and gray.

The Criollo typically has a slightly concave profile, broad chest, and heavily muscled back, shoulders, and neck.

The Criollo, or Crioulo, traces its ancestry back to the horses Spanish conquistadors brought to South America in the 1500s. Many of these Spanish horses were set free in the wild. Over time, these horses evolved into considerably hardier animals that could survive extreme temperatures and subsist with little water and grass. Today the Criollo breed is prized for its endurance and stamina.

MANGALARGA MARCHADOR

Herd of Mangalarga Marchadors

The Mangalarga Marchador breed originated in Brazil when the foundation stallion Sublime was crossed with local mares of Spanish Jennet and Barb descent. The resulting offspring had a smooth, rhythmic gait—the *marcha*. The Mangalarga Marchador has been selectively bred in Brazil for more than 180 years. The first breed association was formed in Brazil in 1949 to promote the Mangalarga Marchador. Today the Mangalarga Marchador is the national horse of Brazil and is known for its smooth gait, gentle temperament, intelligence, stamina, and comfortable ride.

Mangalarga Marchadors are used on farms and cattle ranches and for endurance riding, trail and pleasure riding, cutting, jumping, and polo.

Mangalarga Marchadors have triangular heads with a straight profile, large nostrils, ears pointing slightly inwards and large, expressive eyes.

CHAPTER 4:

Horses of Europe

SHIRE

The world's largest horse breed is the Shire. It descends from Britain's "great horse," which carried men in heavy armor into battle. The Shire proved just as useful in agriculture and industry as it did in war. Throughout its history, the Shire has been used in agriculture and industry, prized for its immense pulling capacity. Shires are usually bay, brown, black, or gray and have feathered hair below the knee.

The Shire breed gets its name from the shires (counties) of England where the breed was found and developed.

BREED	Shire
ORIGIN	England
COLOR	Black, brown, bay, gray
HEIGHT	At least 17 hands (65 inches) for stallions; 16 hands (64 inches) for mares; 16.2 hands (64.8 inches) for geldings

Shire stallions average just over 17 hands (68 inches) in height and weigh as much as 2,200 pounds.

Before modern machinery, Shire horses were used to plow fields and perform other heavy farm work.

Since the early 1800s, Shire horses have been used to pull carts of ale from breweries to public houses. A few breweries in Great Britain continue this tradition.

SUFFOLK PUNCH

All Suffolk Punch horses are chestnut, with colors ranging from light golden to dark liver.

The Suffolk Punch, also known as the Suffolk horse, is a draft horse. Farmers in Suffolk County in eastern England developed the breed. They are known as strong and faithful workers. Today the breed is used for forestry, other draft work, advertising, and crossbreeding to produce heavy sport horses. Suffolk Punch horses are always chestnut in color.

BREED	Suffolk Punch
ORIGIN	England
COLOR	Chestnut
HEIGHT	16.1–17.2 hands (64.4–68.8 inches)
WEIGHT	1,980–2,200 pounds

CLEVELAND BAY

The Queen of England's Royal Mews (stables) uses Cleveland Bays in ceremonial duties. Here, two Cleveland Bay horses pull a coach near Buckingham Palace.

Suffolk Punch mare and foal

The Cleveland Bay originated in Britain, in the Cleveland area of Northern Yorkshire. It is the oldest established English horse breed. Cleveland Bays are generally bay colored and stand up to 17 hands (68 inches) tall. They are used for carriage driving and riding.

NEW FOREST PONY

The New Forest pony is one of the recognized breeds of mountain and moorland ponies of the British Isles. The breed is native to the New Forest in southern England, where ponies have lived for thousands of years. New Forest ponies are known for their intelligence, strength, versatility, surefootedness, gentleness, and eager-to-please temperament.

New Forest pony mare and foal

Children and adults can ride New Forest ponies. The pony breed is used for show jumping, dressage, driving, and cross-country events, among other equestrian disciplines.

New Forest ponies typically stand around 12–14.2 hands (48–56.8 inches) tall. Bay, brown, chestnut, and gray are the most common colors.

BREED	**New Forest pony**
ORIGIN	**England**
COLOR	**Any except for piebald, skewbald, spotted, or blue-eyed cream**
HEIGHT	**Up to 14.2 hands (56.8 inches)**

The New Forest pony gets its name from the New Forest in southern England. The New Forest is one of the largest remaining areas of unenclosed land in southern England.

Free-roaming New Forest pony

Buckskin New Forest pony

New Forest ponies still roam the heaths, woodlands, and bogs of the New Forest as they have for centuries. The ponies are sometimes described as 'the architects of the Forest' because their grazing creates the close-cropped lawns between the wooded areas.

The ponies living in the New Forest are not completely feral. Each pony is owned by a local person whose property has common rights allowing them to turn out ponies to graze on the Forest.

EXMOOR PONY

The Exmoor is one of the British Isles' mountain and moorland pony breeds.

The Exmoor pony is a rare breed native to the British Isles. The ponies have roamed southwestern England for centuries. The breed takes its name from the Exmoor—the open, hilly moorland in Somerset and Devon, England—where some ponies still live semi-feral. The Exmoor pony has a strong, powerful build and is noted for its hardiness and endurance.

All Exmoor ponies are some shade of brown with darker legs and mealy oatmeal-colored markings around the muzzles, eyes, and sometimes under the belly.

Exmoor ponies are a common sight in Exmoor National Park, where a number of managed herds graze and roam freely. The moorland provides a varied diet of grasses, rushes, heather, and gorse.

Exmoor ponies are exposed to wet winters with cold temperatures and heavy winds. The ponies grow a thick, two-layered coat to protect against harsh winter conditions. Their summer coat is fine and glossy.

Adult Exmoor ponies don't vary much in size. The average Exmoor stands about 12.2 hands (48.8 inches) tall. There is also relatively little variation in color and markings.

Exmoor foals are born with mealy markings set against a much lighter coat color than adults. This changes as they grow.

FELL PONY

The Fell pony gets its name from the fells (mountains or hills) of northern England where the breed originated. It has been a recognizable breed since Roman times in England. The Fell pony is noted for its hardiness, agility, strength, and adaptability. Fell ponies were once used as packhorses, carrying copper, slate, iron, and lead. Today they are used for pleasure riding, driving, and as a family pet.

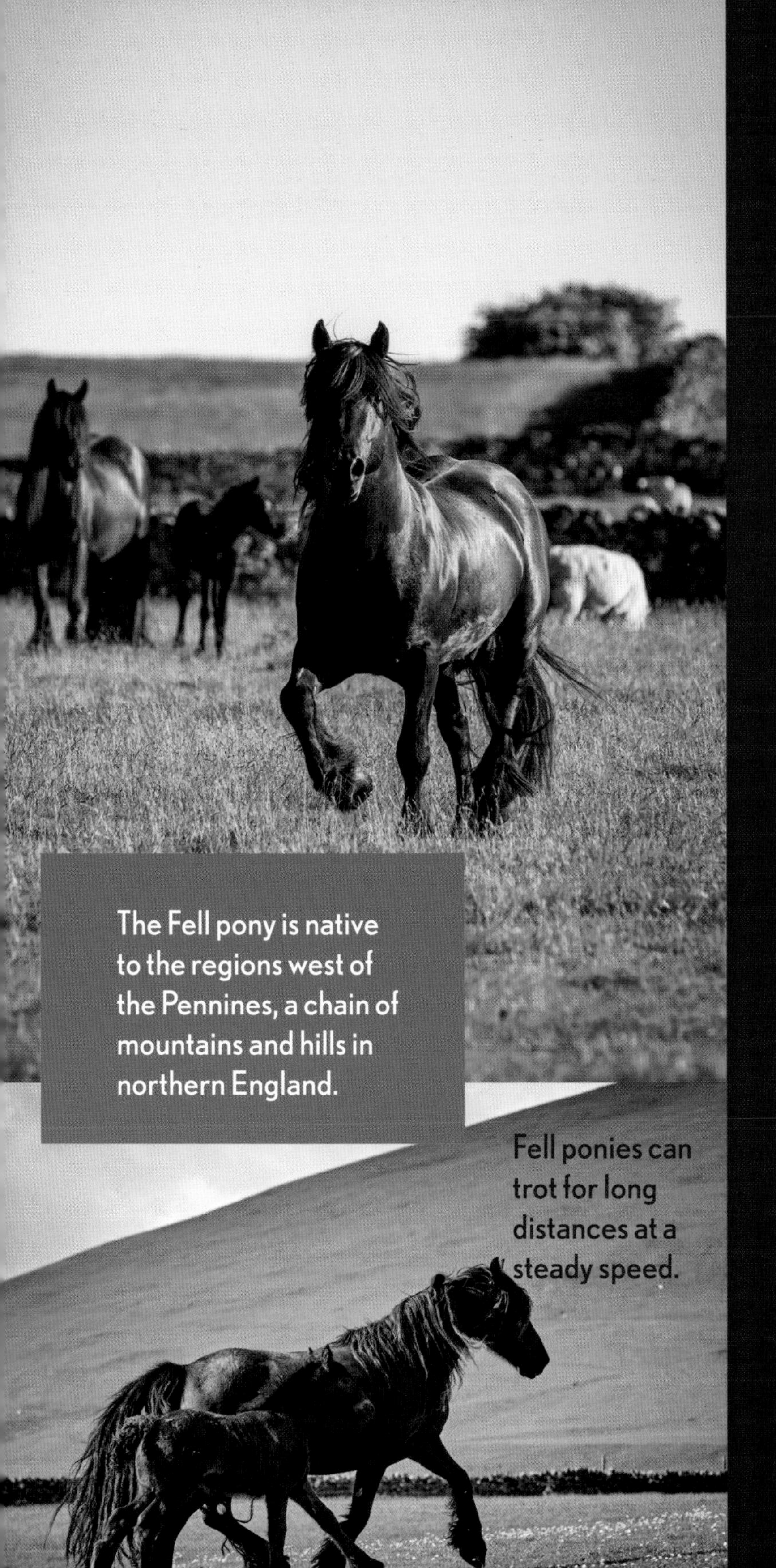

The Fell pony is native to the regions west of the Pennines, a chain of mountains and hills in northern England.

Fell ponies can trot for long distances at a steady speed.

BREED	Fell pony
ORIGIN	England
COLOR	Bay, black, brown, gray
HEIGHT	13–14 hands (52–56 inches)

DARTMOOR PONY

Although ponies roam freely on the moor at Dartmoor National Park, each belongs to an owner. The owners round up their ponies every year and decide which ponies to keep on the moor and which to sell.

The Dartmoor pony is named for the barren Dartmoor in Devon, England, where the breed has lived for centuries. One of the earliest references to the Dartmoor pony is in the will of Saxon Bishop Awlfwold of Crediton, who died in 1012. Many Dartmoor ponies were used to carry tin from mines to surrounding towns. When mines closed, most Dartmoor ponies were let loose to roam the moor, except for those kept by local farmers for use on the farms. Today many ponies live semi-feral on Dartmoor.

BREED	**Dartmoor pony**
ORIGIN	**England**
COLOR	**Bay, black, brown, chestnut, gray, roan**
HEIGHT	**Up to 12.2 hands (48.8 inches)**

The area from which the Dartmoor pony originates is rocky, barren moorland. The sparse grazing and harsh weather ensures only the toughest inhabitants thrive. Because of the extreme conditions, the Dartmoor is a particularly hardy breed.

Because of their calm temperament and surefootedness, Dartmoor ponies can make excellent riding ponies for children. They are also used as foundation stock for breeding larger ponies. The Dartmoor Pony Society, which set the breed standard, only accepts solid-colored ponies.

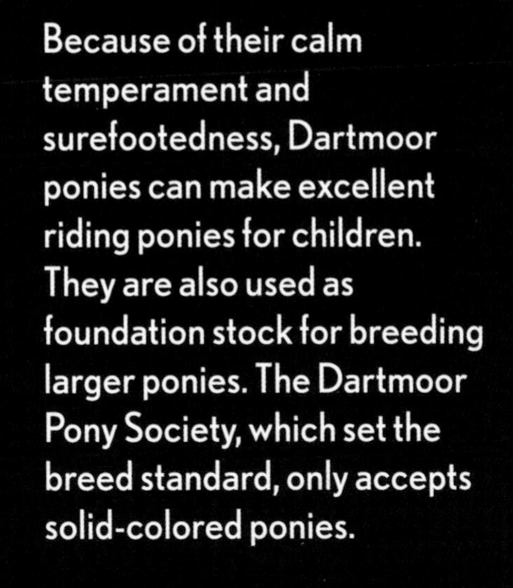

WELSH PONIES

The Welsh mountain pony originated in the mountains of Wales where conditions were harsh and vegetation was sparse. Bands of ponies roamed in a semi-feral state, climbing mountains, jumping ravines, and running over rugged terrain. They developed into a hardy breed.

The Welsh pony and cob is a group consisting of four types, called sections—the Welsh mountain pony (section A), the Welsh pony (section B), the Welsh pony of cob type (section C), and the Welsh cob (section D). Welsh ponies and cobs in all sections are known for their hardiness, intelligence, friendliness, and even temperaments. They originated in the mountains of Wales with a lineage that predates the Roman Empire. Throughout history, Welsh ponies and cobs have been used as warhorses, pit ponies in mines, and working animals on farms. Today they are used for nearly every competitive discipline as well as for pleasure and tail riding.

BREED	**Welsh mountain pony (section A)**
ORIGIN	**Wales**
COLOR	**Any except piebald or skewbald**
HEIGHT	**Up to 12 hands (48 inches) in U.K.; 12.2 hands (48.8 inches) in U.S.**

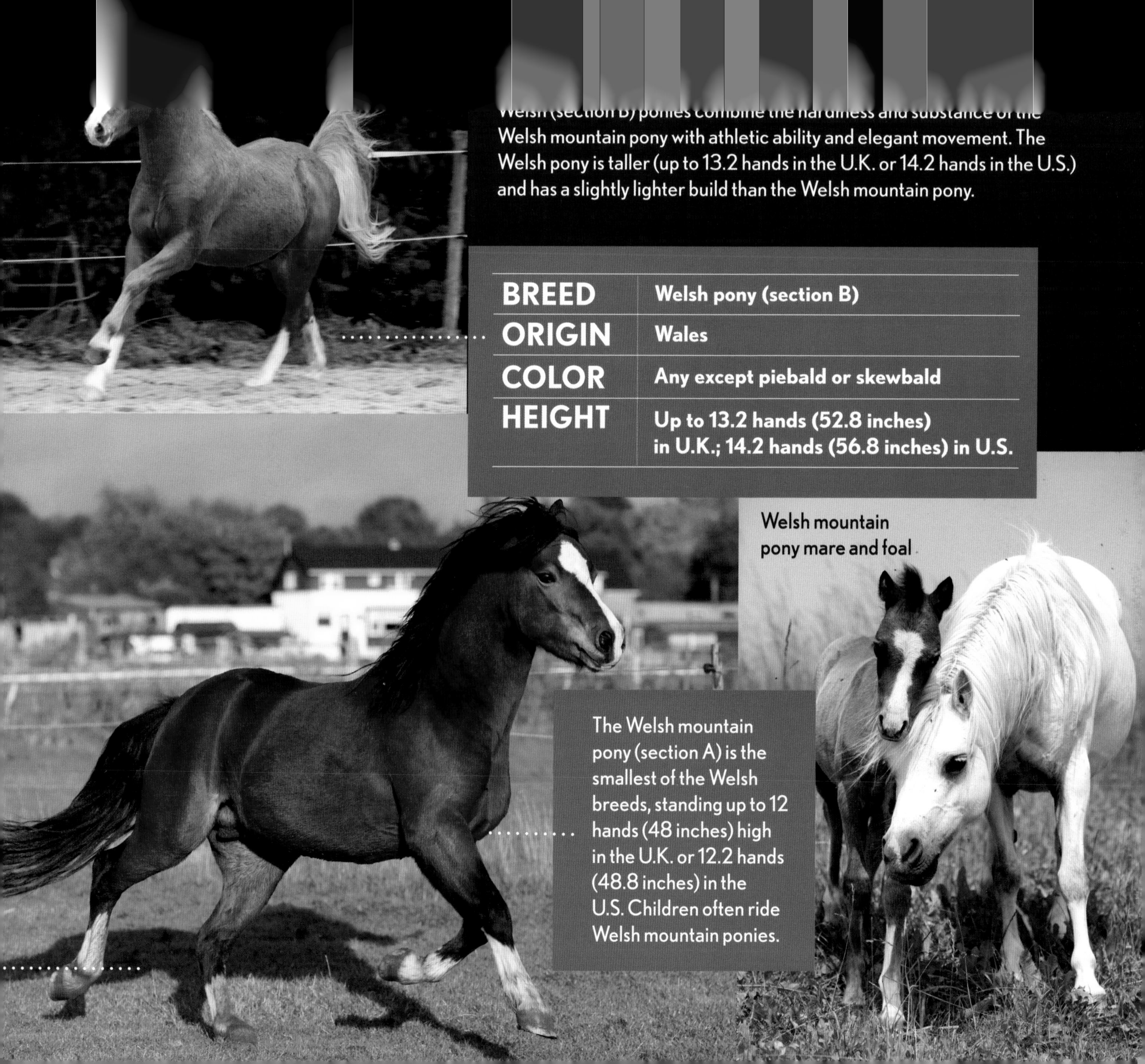

Welsh (section B) ponies combine the hardiness and substance of the Welsh mountain pony with athletic ability and elegant movement. The Welsh pony is taller (up to 13.2 hands in the U.K. or 14.2 hands in the U.S.) and has a slightly lighter build than the Welsh mountain pony.

BREED	**Welsh pony (section B)**
ORIGIN	**Wales**
COLOR	**Any except piebald or skewbald**
HEIGHT	**Up to 13.2 hands (52.8 inches) in U.K.; 14.2 hands (56.8 inches) in U.S.**

Welsh mountain pony mare and foal

The Welsh mountain pony (section A) is the smallest of the Welsh breeds, standing up to 12 hands (48 inches) high in the U.K. or 12.2 hands (48.8 inches) in the U.S. Children often ride Welsh mountain ponies.

WELSH COBS

The Welsh pony of cob type (section C) was first developed by crossbreeding a Welsh mountain pony (section A) and a Welsh cob (section D). This is the rarest of the four sections.

BREED	Welsh pony of cob type (section C)
ORIGIN	Wales
COLOR	Any except piebald or skewbald
HEIGHT	Up to 13.2 hands (52.8 inches)

BREED	**Welsh cob (section D)**
ORIGIN	**Wales**
COLOR	**Any except piebald or skewbald**
HEIGHT	**At least 13.2 hands (52.8 inches)**

The Welsh cob (section D) is the largest of the Welsh pony and cob breed registries. They must be taller than 13.2 hands (52.8 inches), with no upper limit. Welsh cobs compete in a variety of equestrian disciplines.

Welsh cobs embody strength, hardiness, and agility.

CONNEMARA PONY

Connemara ponies come in a variety of colors, but gray, bay, black, and brown are most common.

BREED	**Connemara pony**
ORIGIN	**Ireland**
COLOR	**Bay, black, brown, buckskin, chestnut, cream, dun, gray, palomino, roan**
HEIGHT	**12.2–14.2 hands (48.8–56.8 inches)**

The Connemara is a pony breed native to the Connemara region of County Galway in western Ireland. The breed is known for its hardiness, athleticism, surefootedness, versatility, and good disposition. Connemara ponies are used as general riding ponies for children and adults and as hunters, jumpers, and show ponies.

The Connemara is one of several mountain and moorland pony breeds native to the British Isles.

The Connemara is a versatile breed used for show jumping, dressage, endurance riding, and other events.

IRISH COB

The Irish cob goes by many names, including gypsy cob, gypsy horse, gypsy vanner, and tinker cob. The breed is native to Ireland and has been associated with the Irish Traveller and Romani traveling people of the British Isles. The Irish cob resulted from crossing the English Thoroughbred, Connemara pony, and Irish draught horse. Originally used for pulling carriages, the Irish cob is now used in a variety of disciplines.

Irish cobs are usually piebald (irregular, large areas of black and white). They may also be skewbald (large white patches with any color except black) or any solid color.

Irish cobs are known for their strength, stamina, intelligence, and willingness to please.

BREED	Irish cob
ORIGIN	Ireland
COLOR	Any
HEIGHT	13-16.2 hands (52-64.8 inches)

The long, feathered hair on the legs is considered a characteristic and decorative feature of the Irish cob, but not required for registration.

ERISKAY PONY

Eriskay pony on Beinn Sciathan, the highest hill on the Outer Hebridean island of Eriskay

The Eriskay pony developed in the Hebrides islands of western Scotland. The breed gets its name from the island of Eriskay in the Outer Hebrides. This ancient breed has roots in Celtic and Norse breeding. Though numerous in the past, the Eriskay saw a dramatic decline in their numbers in the 19th century due to increased crossbreeding. Today the Eriskay is rare. Eriskay ponies are used for light draft work, as children's mounts, and for driving.

BREED	**Eriskay pony**
ORIGIN	**Scotland**
COLOR	**Bay, black, gray**
HEIGHT	**12-13.2 hands (48-52.8 inches)**

Over the centuries, the Eriskay evolved into the hardy, versatile, and people friendly pony we see today.

Eriskay ponies have dense, waterproof coats that protect them from harsh weather.

SHETLAND PONY

Shetland ponies are named for the Shetland Islands off the coast of northern Scotland, where they have lived for centuries. Shetland ponies adapted to the harsh climate and scant food supply in their native islands. Shetlands were used to pull plows, carts, and buggies. In the 1840s, Shetland ponies replaced women and children working in Britain's coal mines. Today the Shetland pony is a popular children's mount.

Unlike most other breeds, Shetland ponies are measured in inches rather than hands. Shetland ponies may not exceed 42 inches in height at the withers.

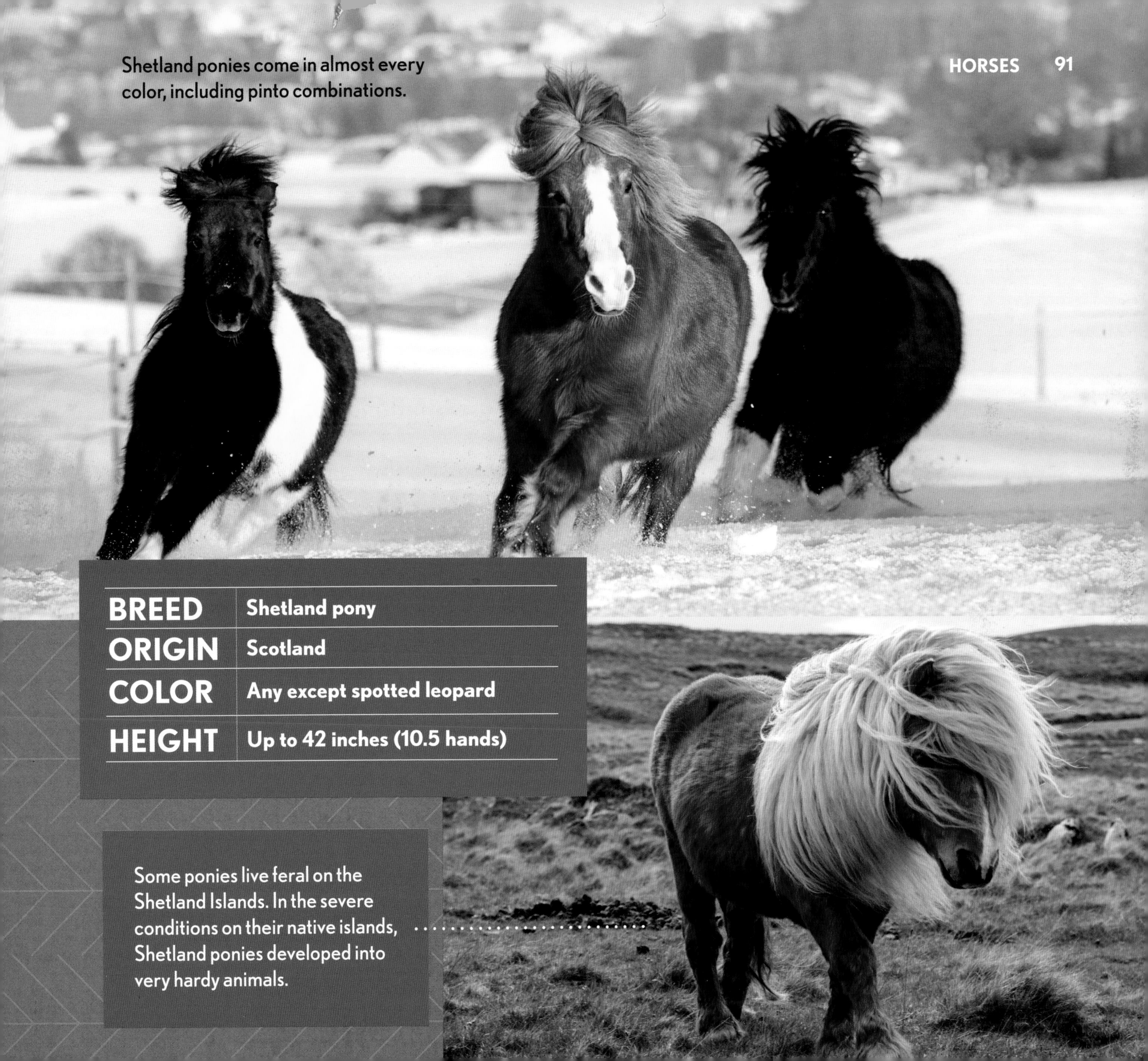

Shetland ponies come in almost every color, including pinto combinations.

BREED	Shetland pony
ORIGIN	Scotland
COLOR	Any except spotted leopard
HEIGHT	Up to 42 inches (10.5 hands)

Some ponies live feral on the Shetland Islands. In the severe conditions on their native islands, Shetland ponies developed into very hardy animals.

CLYDESDALE

Most Clydesdales have white markings on the face and legs.

The Clydesdale is a heavy draft horse known for its large size and long, silky hair on its legs. The breed gets its name from Clydesdale (now called Lanarkshire), Scotland, where it originated. Clydesdales have plowed fields, hauled heavy loads, and served as a mascot of various beer brands, including Budweiser. Modern Clydesdales stand up to 18 hands high and weigh up to 2,200 pounds.

BREED	Clydesdale
ORIGIN	Scotland
COLOR	Bay, black dark brown, chestnut, gray, roan
HEIGHT	16–18 hands (64–72 inches)

Clydesdales are used for both riding and driving. They are frequently crossbred with Thoroughbreds to develop strong, sensible sports horses.

The Budweiser Clydesdales are some of the most famous Clydesdales. Teams of these horses frequently appear in parades and fairs.

FRIESIAN

BREED	Friesian
ORIGIN	Netherlands
COLOR	Black
HEIGHT	14.2–17 hands (56.8–68 inches)

Friesian horses are always black, though the color can range from very dark brown to true black.

The Friesian horse is native to the province of Friesland in the northern Netherlands. It is one of Europe's oldest breeds. Friesians were once used to carry knights into battle. The breed nearly became extinct before World War I, but has since been revived. Today its numbers and popularity are growing.

Friesian horses have a long mane and tail that are often wavy.

Friesian foal

FRIESIAN

The Friesian is one of the best carriage horses. Throughout Europe, royal courts used Friesians as coach horses. Today Friesians are often seen in carriage events.

Friesians are known for their high-stepping trot. Friesians were often used for racing short distances.

The average Friesian stands about 15.3 hands (61.2 inches) high.

The Friesian horse is used in harness and under saddle, particularly in dressage.

DUTCH WARMBLOOD

The Dutch warmblood evolved from two native Dutch breeds—the Gelderlander and the Groningen. As its name implies, this is a breed of warmblood horse. Warmbloods are typically athletic, agile horses known for their trainability and calm temperament. Dutch warmbloods are often used in equestrian competition, including dressage and show jumping.

Dutch warmbloods are bred to perform in dressage and show jumping.

BREED	Dutch warmblood
ORIGIN	Netherlands
COLOR	Bay, black, brown, chestnut, gray
HEIGHT	At least 15.2 hands (60.8 inches)

The average Dutch warmblood stands about 16.2 hands (64.8 inches) tall.

Dutch warmbloods can be black, brown, bay, chestnut, or gray. White markings on the face or legs are common.

FINNHORSE

The Finnhorse is a descendent of the northern European domestic horse with both warmblood and heavier draft influences. The Finnhorse has been used for everything from agriculture and forestry to harness racing and riding. Today there are four types of Finnhorse—a lightly built trotter, a heavy draught (draft) horse, a versatile riding horse, and a smaller pony-sized horse.

Chestnut is the most common color of Finnhorse.

This versatile breed is often called the "Finnish Universal" in its native country because it fulfills all of Finland's horse needs.

The Finnhorse is often used for harness racing. Its popularity for harness racing has helped the Finnhorse breed survive.

BREED	**Finnhorse**
ORIGIN	**Finland**
COLOR	**Chestnut, bay, gray, brown, black, palomino**
HEIGHT	**14.2-15.3 hands (56.8-61.2 inches)**

GOTLAND PONY

The Gotland pony is a primitive breed native to the island of Gotland in Sweden, where ponies of this type have been documented as far back as the Stone Age. They are also known as Russ or Skogruss (meaning "little horse of the woods"). Gotland ponies are hardy, athletic, and docile with a gentle disposition. The Gotland is a popular riding pony for children and is often used for driving, racing, and farm work.

Gotland ponies have lived on the island's wooded moors for thousands of years.

Dun and bay are the two most common colors of Gotland pony, but all colors except albinos, roans, and piebalds are allowed.

BREED	**Gotland pony**
ORIGIN	**Sweden**
COLOR	**Any except albino, roan, or piebald**
HEIGHT	**11.2-12 hands (44.8-52 inches)**

Gotland ponies faced near extinction by the early 20th century, but locals intervened to save the breed. Thanks to their efforts, Gotland ponies still live on the island for which they are named.

NORWEGIAN FJORD

The Norwegian fjord horse is one of the world's oldest horse breeds. It is believed to have been in western Norway for more than 4,000 years and domesticated as early as 2000 B.C. The fjord horse is strong enough to plow fields and haul timber, yet light and agile enough to make an excellent riding horse. Fjord horses are known for their vigor, stamina, willingness to work, and gentle temperament.

All fjord horses are dun colored. The breed standard recognizes five shade variations. About 90 percent of all fjord horses are brown dun.

BREED	Norwegian fjord
ORIGIN	Norway
COLOR	Dun
HEIGHT	13.2-15 hands (52.8-60 inches)
WEIGHT	900-1,200 pounds

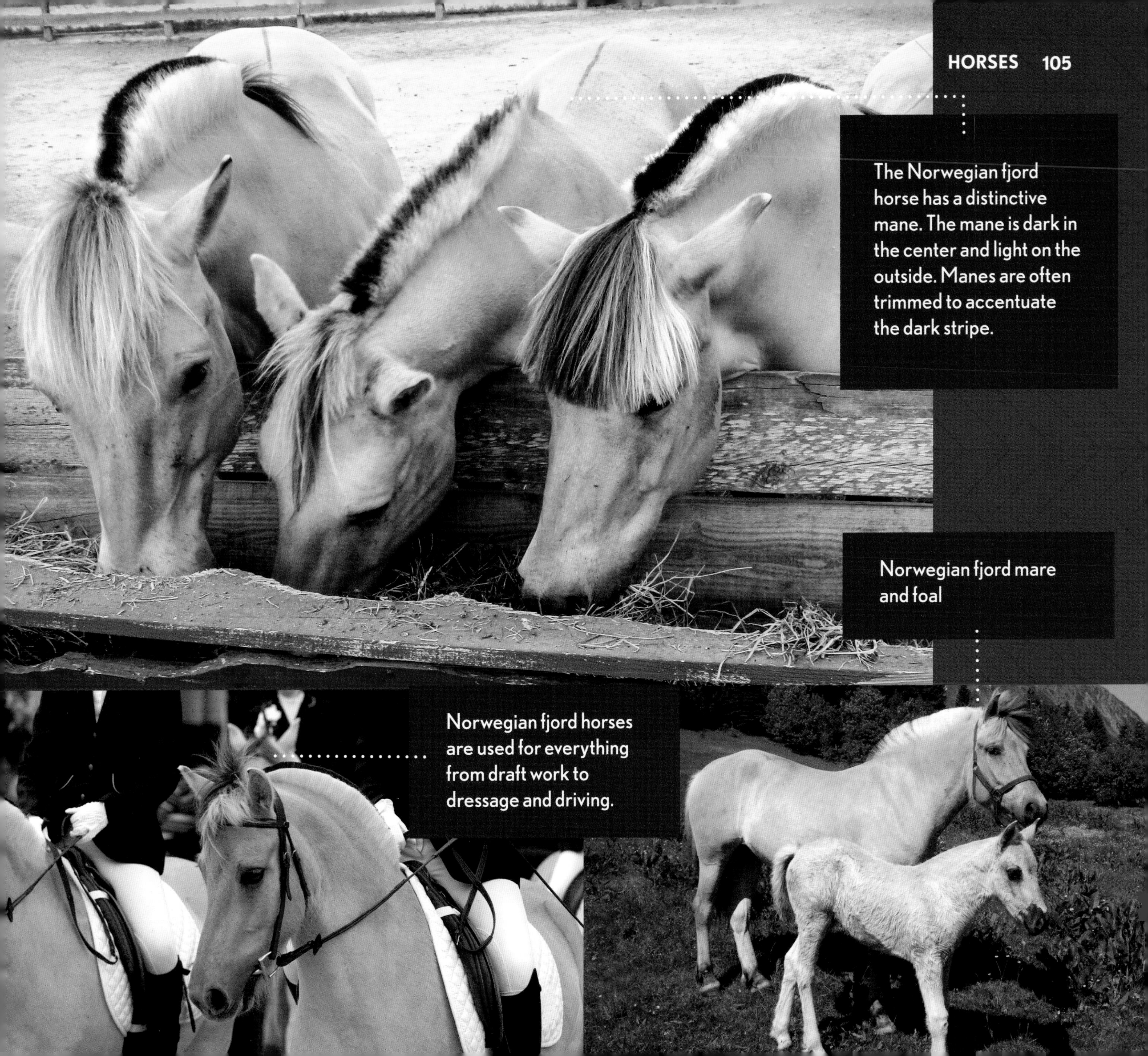

The Norwegian fjord horse has a distinctive mane. The mane is dark in the center and light on the outside. Manes are often trimmed to accentuate the dark stripe.

Norwegian fjord mare and foal

Norwegian fjord horses are used for everything from draft work to dressage and driving.

ICELANDIC

In addition to the usual walk, trot, and canter/gallop gaits, Icelandic horses have two additional gaits—a running walk known as the tölt and a fast flying pace used for racing.

The Icelandic horse developed in Iceland from ponies Norse settlers brought to the island in the 9th and 10th centuries. Although Icelandics are often pony-sized, they are always referred to as horses. Though small, Icelandic horses can carry heavy adult riders. The Icelandic is a versatile riding horse known for its sturdy build, intelligence, and enthusiasm. Icelandic horses are used for traditional sheepherding as well as for pleasure riding, racing, showing, and sport competitions.

BREED	Icelandic
ORIGIN	Iceland
COLOR	Any
HEIGHT	12-14.2 hands (48-56.8 inches)

Icelandic horses come in all coat colors and markings. Chestnut, black, and bay are the most dominant colors.

Young Icelandic horses typically live outside with the herd for their first four years of life. Icelandic horses mature slowly and are not ridden until at least age four or five.

BELGIAN DRAFT

Belgian draft horses are docile and willing workers known for their tremendous weight-pulling capacity.

The Belgian draft horse descended from the Flemish "great horse," the medieval warhorse that carried armored knights into battle. The breed, which is also known as the Belgian heavy horse, is from the Brabant region of Belgium. Belgians have thick, muscular bodies and short legs. Belgians are still used as working animals today, but are also used for pleasure riding, pulling wagons, and as show horses.

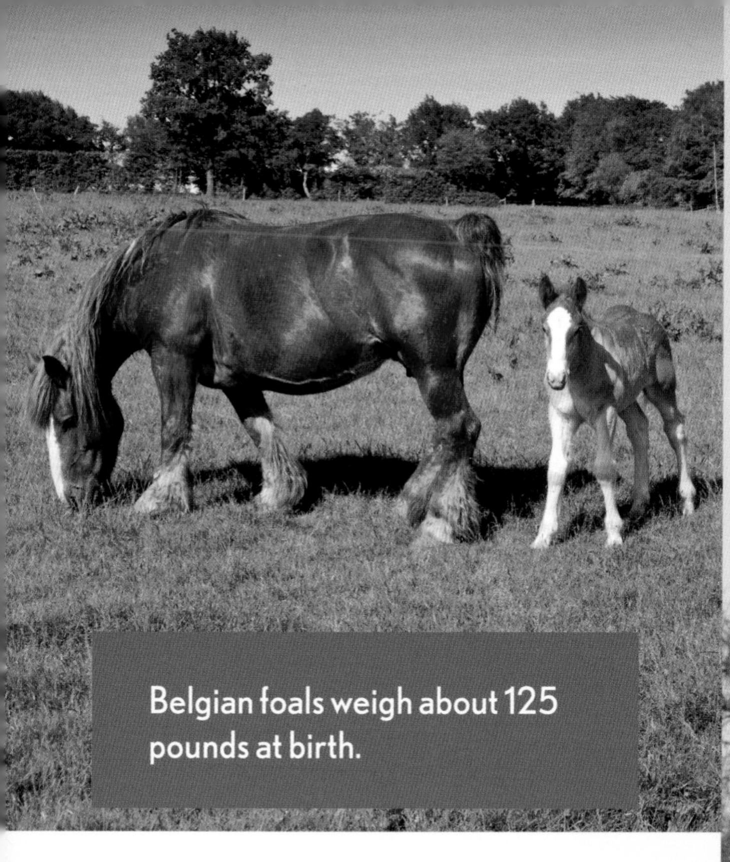

Belgian foals weigh about 125 pounds at birth.

The average Belgian draft horse stands 16.2–17 hands (64.8–68 inches) tall.

BREED	**Belgian draft**
ORIGIN	**Belgium**
COLOR	**Bay, chestnut, sorrel, gray, roan brown, black**
HEIGHT	**16–18 hands (64–72 inches)**
WEIGHT	**1,800-2,200 pounds**

Today's Belgian draft horse is often chestnut-sorrel or roan with a blonde or white mane and tail.

CAMARGUE

Camargue horses galloping through water is a popular image associated with the region.

The Camargue horse is an ancient breed native to the Camargue area in southern France. They have lived for centuries in the wetlands of the Camargue—the region between the Mediterranean Sea and two arms of the Rhône River delta. While some live semi-feral in the *Parc naturel régional de Camargue* (Regional Nature Park of the Camargue), most Camargue horses are used for riding. The Camargue horse is known for its hardiness, stamina, agility, intelligence, and calm temperament.

BREED	Camargue
ORIGIN	France
COLOR	Gray
HEIGHT	13.1–14.3 hands (52.4–57.2 inches)
WEIGHT	770–1,100 pounds

The Camargue horse is the traditional mount used by Camargue cowboys (*gardians*) who herd the area's distinctive black bulls.

Camargue horses are born with a black or brown coat that turns light gray as they age.

Camargue horses have lived in the wetlands of the Camargue region for hundreds of years and have adapted to the harsh conditions.

The Camargue is a small horse with a short neck, deep chest, compact body, and strong, well-jointed limbs.

Camargue horses provide tourists with the opportunity to explore the area on horseback.

Many horses of the Camargue roam freely in the marshlands in small herds.

PERCHERON

Before mechanization revolutionized farming, Percherons were widely used in agriculture. Today they are still used to pull heavy loads.

The Percheron is a heavy draft horse breed that originated in the Perche region in northwestern France. Once a warhorse, then a popular draft horse, the Percheron today is used for everything from parades and carriage rides to competition hitching, halter, and riding classes. Percherons are prized for their agility, energy, endurance, calm disposition, and willingness to please.

BREED	**Percheron**
ORIGIN	**France**
COLOR	**Black, gray, chestnut, bay, roan, sorrel**
HEIGHT	**15–19 hands (60–76 inches)**

Percheron size varies by country. In its native France, Percherons generally stand 15.1–18.1 hands tall and weigh 1,100–2,600 pounds.

Percherons are often crossed with Thoroughbreds, warmbloods, and Spanish breeds to create sport horses for dressage, hunting, and pleasure riding.

BOULONNAIS

Boulonnais mare and foal

The Boulonnais is a draft horse breed that developed in northern France. The breed's history begins before the Crusades. Over the years, Spanish Barb, Arabian, and Andalusian blood have been added to the breed. The modern Boulonnais is an energetic, elegant draft horse that has contributed to other breeds, including the Italian heavy draft, the Ardennes, and the Schleswig horse.

Boulonnais horses are usually gray, ranging from very light to dark dappled gray.

BRETON

The Breton is a draft horse developed in Brittany, a province in northwest France. There are three types of Breton. The Corlay or Central Mountain Breton is the smallest, standing 14.3–15.1 hands high. The Postier Breton stands an average of 15.1 hands high and is used as a coach horse and for light farm work. The heavy draft Breton is the largest, standing 15.2–16.2 hands tall.

Bretons usually have a chestnut coat, often with a flaxen mane and tail. Bretons can also have bay, gray, or roan coats.

Breton mare and foal

HAFLINGER

Haflingers are always chestnut, with shades ranging from a pale gold to a rich red-gold. The mane and tail are flaxen or white.

The Haflinger (also known as the Avelignese) developed in the mountains of Austria and northern Italy. The breed gets its name from the village of Hafling in the South Tyrol region of northern Italy. The hardy Haflinger was traditionally used for farm labor, pulling carts and carriages, transportation, and pack hauling. Today Haflingers are also used in various equestrian disciplines.

BREED	Haflinger
ORIGIN	Austria, Italy
COLOR	Chestnut
HEIGHT	13.2-15 hands (52.8-60 inches)

Haflingers were developed in the eastern Alps of present-day Austria and northern Italy. Haflingers needed to be agile and surefooted to traverse narrow mountain paths.

Haflingers are now used for Western and trail riding, endurance riding, dressage and jumping, vaulting, and therapeutic riding programs.

LIPIZZAN

BREED	Lippizan
ORIGIN	Austria, Slovenia
COLOR	Gray, bay, brown, black
HEIGHT	14.2–16.1 hands (56.8–64.4 inches)

Lipizzan horses are famous for the difficult "airs above the ground" dressage movements.

The Lipizzan (or Lipizzaner) was developed in the 16th century by the Habsburg nobility and the Austro-Hungarian Monarchy. The breed was named after the stud farm founded in 1580 at Lipizza, in present-day Slovenia, which was then part of the Austro-Hungarian Empire. Lipizzaners are closely associated with the Spanish Riding School of Vienna, Austria, where they are trained in classical dressage. This small but powerful breed is known for its intelligence, beauty, and gracefulness.

Lipizzaner foals are born with a black or bay coat that turns gray as they age.

Lipizzaners were originally all colors, but today are usually gray. Bay and brown occur rarely.

HANOVERIAN

The Hanoverian is a warmblood horse originating in Lower Saxony, northern Germany, the former kingdom of Hanover, where a flourishing horse-breeding industry has existed for hundreds of years. Hanoverians have been used as carriage horses, in agriculture, and for military service. Today's Hanoverians excel in competitive equestrian sports. The modern Hanoverian is known for its athleticism, agility, speed, beauty, and grace.

Hanoverians can be any solid color. Cremello, palomino, or buckskin coats cannot be registered.

The Hanoverian has light and elegant gaits characterized by a floating trot, a ground-covering walk, and a round, rhythmic canter.

BREED	Hanoverian
ORIGIN	Germany
COLOR	Chestnut, bay, brown, black, gray
HEIGHT	15-18 hands (60-72 inches)

HOLSTEINER

BREED	Holsteiner
ORIGIN	Germany
COLOR	Bay, brown, gray, chestnut, black
HEIGHT	16-17 hands (64-68 inches)

Although the population is relatively small, Holsteiners are a dominant force in show jumping.

The Holsteiner (or Holstein) has been bred in the Schleswig-Holstein region of northern Germany since the 13th century. Over the years, Holsteiners have served as strong and steady farm workers, courageous and capable warhorses, and high-stepping carriage horses. Thoroughbred blood introduced after World War II added refinement and jumping ability. Today the Holsteiner is a leading German sport horse, particularly suited for jumping, dressage, driving, and eventing (where a rider/horse competes across three disciplines: dressage, cross-country, and show jumping).

All solid coat colors are accepted in Holsteiner breed registries, but bay shades are most common.

Holsteiners grazing

OLDENBURG

The Oldenburg is well suited for show jumping, dressage, hunter classes, and eventing because of its temperament, character, and rideability.

The Oldenburg (or Oldenburger) originated in northern Germany, in what was the Grand Duchy of Oldenburg. The breed can be traced back to the 17th century, with bloodlines based on the Friesian horse. Breeders have repeatedly introduced new bloodlines to adapt the Oldenburg to suit market needs. Oldenburgs have served as farm horses, carriage horses, artillery horses, and all-purpose riding horses. Today Oldenburgs are bred in order to produce superior sport horses.

BREED	Oldenburg
ORIGIN	Germany
COLOR	Black, brown, bay, chestnut, gray
HEIGHT	16-17.2 hands (64-68.8 inches)

The modern Oldenburg is particularly successful in dressage.

Oldenburgs are known for their expressive, elastic gaits.

LUSITANO

The same qualities that made the Lusitano a great warhorse make it a prized mount in Portuguese bullfighting, where the bull is not killed. Lusitanos must move nimbly around a charging bull and remain calm.

The Lusitano (also known as Pure Blood Lusitano or Puro Sangue Lusitano) is a Portuguese breed closely related to the Spanish Andalusian. The two breeds, which developed on the Iberian Peninsula, were considered one breed called the Andalusian until the late 1960s. When the Portuguese and Spanish studbooks split in 1966, the Portuguese strain was named the Lusitano, after the word *Lusitania*, the ancient Roman name for the region that present-day Portugal occupies. Lusitanos are known for their intelligence, calm temperament, and willing nature.

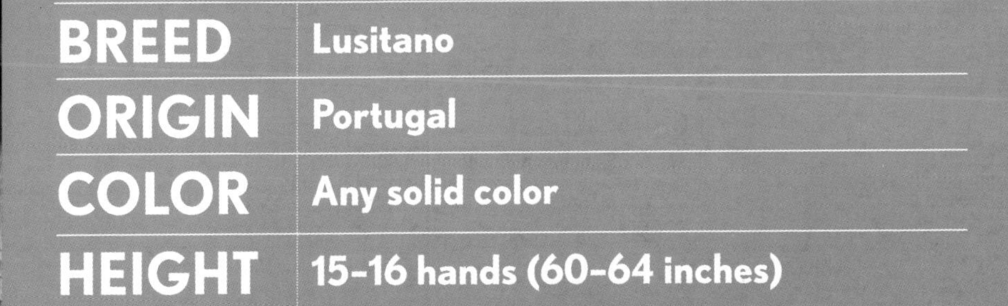

BREED	Lusitano
ORIGIN	Portugal
COLOR	Any solid color
HEIGHT	15–16 hands (60–64 inches)

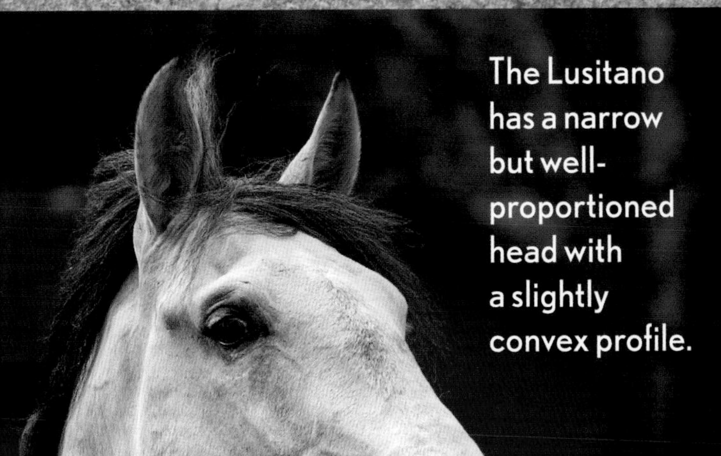

The average Lusitano stands 15.1–15.3 hands (60.4–61.2 inches) tall.

Lusitanos are usually bay, gray, or chestnut, although they can be any solid color, including dun, palomino, and black.

The Lusitano has a narrow but well-proportioned head with a slightly convex profile.

ANDALUSIAN

The Andalusian excels at the equestrian disciplines of dressage, show jumping, and driving.

The Andalusian originated in and derived its name from the Spanish province of Andalusia. Its ancestors are the Iberian horses of Spain and Portugal. The Andalusian, also known as the Pure Spanish Horse or Pura Raza Española, has influenced many other breeds around the world, including the Lipizzaner and the Peruvian Paso. Over its centuries of development, the Andalusian has been bred for athleticism and stamina. Modern Andalusians are used for dressage, show jumping, driving, and bullfighting. They are known for their intelligence, docile temperament, and sensitivity.

BREED	Andalusian
ORIGIN	Spain
COLOR	Gray, white, bay, black, dun, palomino, chestnut
HEIGHT	15–16.2 hands (60–64.8 inches)

The Carthusian is a sub-strain of the breed, which many breeders consider the purest strain of Andalusian.

Most Andalusians today are gray or white. In the past, all coat colors were found, including spotted.

TRAKEHNER

Trakehner mare and foal

The Trakehner is one of Europe's oldest warmblood breeds. The breed developed at a stud farm King Frederick Wilhelm I of Prussia established in Trakehnen, East Prussia, in 1732. A small, native breed called the Schwaike was crossed with English Thoroughbreds and Arabians. The result was a surefooted, intelligent, athletic horse. Today the Trakehner is considered one of the most elegant European warmbloods, prized for its jumping and dressage talents.

BREED	Trakehner
ORIGIN	East Prussia
COLOR	Any
HEIGHT	15.2–17 hands (60.8–68 inches)

Trakehners make excellent show jumpers because of their powerful hindquarters and strong joints and muscles.

Trakehner stallion

The Trakehner excels at dressage because of its sensitivity, intelligence, and elegant gaits—particularly its light, floating trot and soft, balanced canter.

ORLOV TROTTER

The Orlov trotter helped introduce harness racing to Russia. At one time it was the fastest harness horse in Europe.

The Orlov trotter is Russia's most famous horse breed. The breed was developed in the late 18th century by Count A. G. Orlov, who crossed various European mares with Arabian stallions. Orlov trotters were used for riding, harness racing, carriages, and to improve other Russian breeds. Throughout most of the 19th century, no other trotting breed could match the Orlov trotter's speed, stamina, and hardiness. The breed started to decline in the 20th century but dedicated enthusiasts are working to preserve the Orlov trotter.

Many Orlov trotters are gray. Gray horses are born dark and lighten as they age.

BREED	Orlov trotter
ORIGIN	Russia
COLOR	Gray, white, black, bay, chestnut
HEIGHT	15.2–17 hands (60.8–68 inches)

Orlov trotters have big heads with large, expressive eyes.

RUSSIAN DON

The Don played an important role in the development of other Russian breeds, including the Budenny (or Budyonnny), shown above.

The Don horse developed in the steppes of southern Russia where the Don River flows. Don horses have been bred there since the 1700s. The Don horse was the war mount of the famed and feared Cossack Cavalry, who helped drive out Napoleon's invading troops. Today the Russian Don is used under saddle and in harness for recreational riding, equine tourism, and sport competition. Russian Dons are known for their endurance, strength, and hardiness.

BREED	Russian Don
ORIGIN	Russia
COLOR	Bay, black, brown, chestnut, gray
HEIGHT	15.1–16 hands (60.4–64 inches)

The Don horse developed in the treeless steppes of southern Russia. Chestnut and bay are the most common colors.

Russian Don mare and foal

CHAPTER 5:

Horses of Africa

BARB

Gray is one of the most common colors of Barb horse. Others include bay, black, chestnut, and brown.

The Barb (or Berber) horse is a light riding horse known for its stamina and hardiness. The breed originated in northern Africa. There is some controversy over whether Barb and Arabian horses share a common ancestor, or if the Arabian was a predecessor of the Barb. When the Moors invaded Spain in 711, they brought their Barb horses with them. Barb horses played a large part in the development of other breeds, especially the Andalusian. Today the Barb horse is bred primarily in Morocco, Algeria, Spain, and southern France.

The Barb horse has long been associated with the Berber people of North Africa.

Barb horses are fast, agile, and able to survive desert conditions. They can travel over long distances even when supplies are limited.

BREED	**Barb**
ORIGIN	**North Africa**
COLOR	**Brown, black, gray, bay, sorrel, chestnut, dun, roan, buckskin, grullo**
HEIGHT	**13.2–15 hands (52.8–60 inches)**

BASUTO PONY

The rocky terrain helped develop the Basuto pony into a hardy, surefooted horse with great endurance.

BREED	**Basuto pony**
ORIGIN	**Lesotho, South Africa**
COLOR	**Chestnut, brown, bay, gray, black**
HEIGHT	**Up to 14.2 hands (56.8 inches)**

The Basuto (or Basotho) is a pony breed native to the enclave of Lesotho in South Africa. The Basuto pony developed from the Cape horse of South Africa sometime after 1825. The first horses in South Africa, sent by the Dutch East India Company, arrived in 1653. These original horses became the founders of the Cape horse. The Basuto pony was recognized as a distinct breed by 1870 and was used extensively in the Boer Wars of the late 1800s. Basuto ponies are known for their stamina, docility, surefootedness, courage, and hardiness.

NOOITGEDACHT PONY

The Nooitgedacht pony (or Nooitgedachter) is found in the eastern Transvaal region of South Africa. The breed was developed during the 1950s from the Basuto pony with the addition of some Boer and Arab blood. Nooitgedachters are known for their intelligence, stamina, friendliness, hardiness, and surefootedness. They have fine but strong bone structure, good joints, and excellent hooves that seldom require shoeing. Nooitgedachters are used for show jumping, dressage, polo, endurance riding, Western riding, and on ranches and farms.

Bay, brown, and chestnut roans are the most common colors of Nooitgedacht ponies. Spotted, skewbald, and piebald coats are not permitted by the breed society.

CHAPTER 6:

Horses of Australia

BRUMBY

Brumbies inhabit a variety of habitats, including temperate ranges, alpine and subalpine forests, semiarid plains, rocky ranges, tropical grasslands, and wetlands.

The brumby is a free-roaming feral horse of Australia. Horses arrived in Australia with the First Fleet in 1788. Shipments of working farm horses followed. However, records indicate these escaped or were abandoned in the early 1800s. As machines gradually replaced horses in a range of tasks, many horses were released to join already established feral herds. Today there are an estimated 400,000 brumbies throughout Australia inhabiting a range of habitats.

Brumby in Kosciusko National Park, New South Wales

Brumbies pose a complex management problem because they cause environmental damage and compete with native animals and livestock, but also have cultural and economic value.

AUSTRALIAN STOCK HORSE

The Australian stock horse is known for its stamina, intelligence, strength, and good-natured temperament.

The Australian stock horse is an agile, hardy breed that has been developed specifically to meet the demands of the Australian environment. The history of the breed dates back to 1788 when the first horses arrived in Australia. The weakest were culled and the strongest were bred in order to produce a horse of great stamina and strength. Today the Australian stock horse is used in a variety of riding disciplines as well as by stockmen working with livestock.

Australian stock horse mare and foal

BREED	**Australian stock horse**
ORIGIN	**Australia**
COLOR	**Any**
HEIGHT	**14–16.2 hands (56–64.8 inches)**